How Do Lobsters Stay Young?

and other amazing facts about animal life cycles

by Jillian Powell

Contents

T0187118

Life cycles

Birth to death

All living things, including you, have a life cycle. They start as a seed or egg, grow, **reproduce** and then die. That's their life cycle. Some life cycles are long, some are short, but all are pretty awesome and some are really freaky! This book will introduce you to some amazing facts about animal life cycles.

Strange ...

Some **marine** animals like hydra and sponges can just sprout babies from their body like buds. Now what if your mum could do that!

... but true

A hydra sprouting a bud.

bud

Name: **Charles Darwin**

Dates: 1809–1882

Job: **Brilliant Biologist**

Darwin spent his life studying animals and their life cycles. In the 1830s, he sailed around the world, doing research that led to his famous book *On the Origin of Species*. Darwin was the first to talk and write about 'evolution' – the fact that animals and plants change over time. His research led to lots of discoveries, including the fact that humans are related to apes ... but not everyone liked that idea!

Darwin sailed on the HMS Beagle. It was his job to study the animals and plants he found on the journey.

3

Life spans

Long or short?

Animals have different life spans, meaning some live longer than others. Generally, larger animals live longer than smaller animals. A mouse only lives for two or three years, but an elephant can live to be over 70 years old!

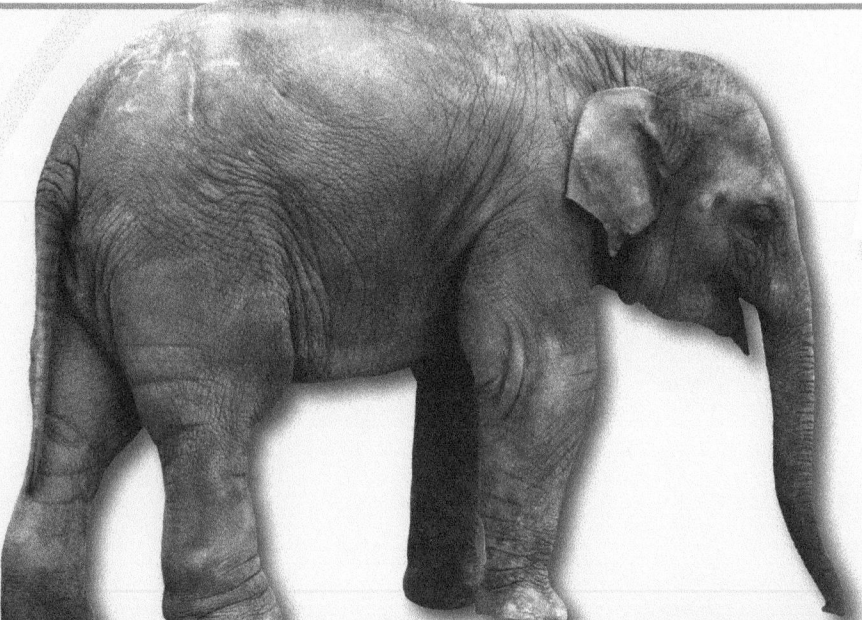

Mice and elephants have very different life spans.

How do lobsters stay young?

Some animals have a few clever tricks to stop them from getting old. Lobsters keep growing new shells and even new limbs as they age*.

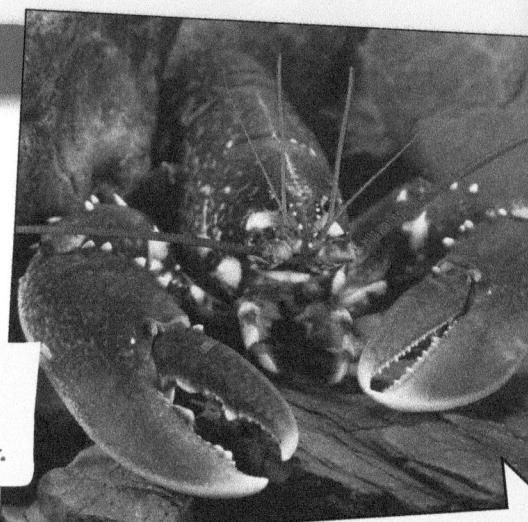

Young lobsters can grow new shells up to 25 times in a year.

*Check out the lobsters on pages 26–27.

Long-life jellies

Some clams can live for 400 years on the seabed. But even they can't beat the jellyfish that can turn back time! Once this jellyfish has had babies, it sinks down to the bottom of the sea and turns into a ball of **cells**. These re-arrange themselves to become a new jellyfish. It can then go on and have more babies. Now *that* is smart!

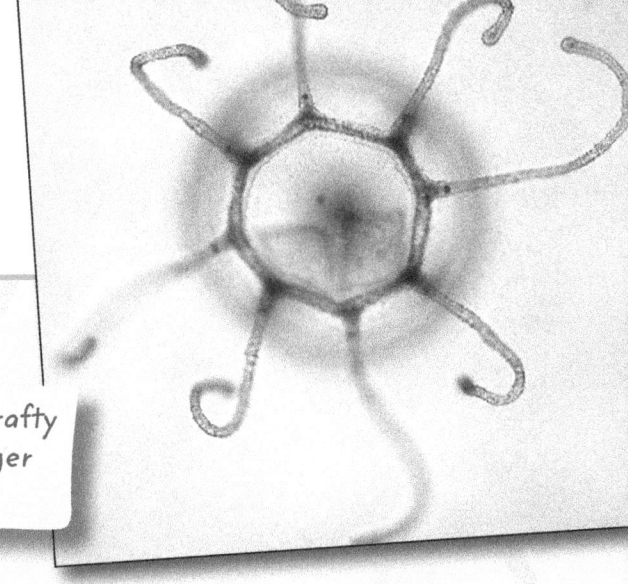

This jellyfish has a crafty way of getting younger rather than older!

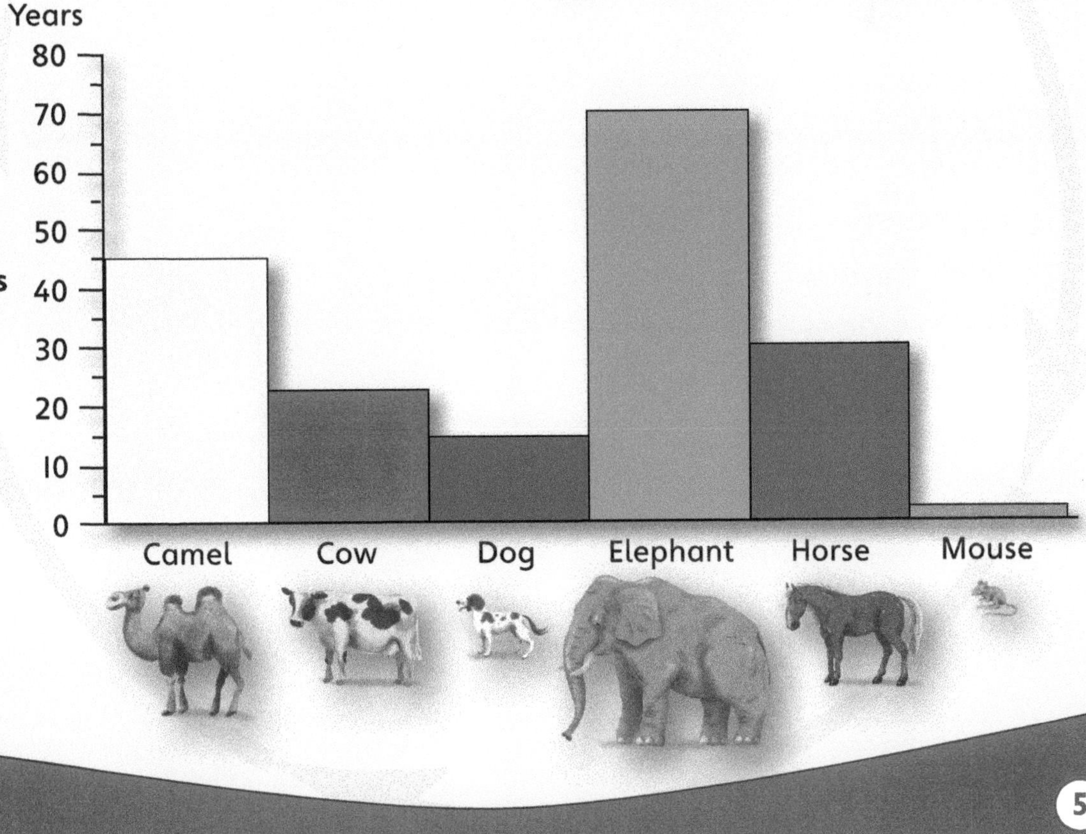

Years
80
70
60
50
40
30
20
10
0
Average life spans
Camel
Cow
Dog
Elephant
Horse
Mouse

Food chains

Recycling energy

During their life cycles, animals need energy to grow, live and reproduce. They get energy from plant or animal foods, and, if they are eaten by another animal, that energy gets recycled. So a **food chain** can form a complete circle of energy.

worm eats the leaves from the trees

frog eats the worm

soil feeds the trees that grow new leaves

HAWK

snake eats the frog

hawk dies and **fungi** help break its body down to feed the soil

hawk eats the snake

Try this!

Can you start your own food chain?

What you need:

- a grassy area
- some small leaves
- some tent pegs
- an onion net

What to do:

1 Find a patch of short damp grass.

2 Count your leaves. Place them on the grass and cover them with the net.

3 Use the tent pegs to hold the net down.

4 Count the leaves daily to see how many disappear. The worms will pull the leaves down to eat them.

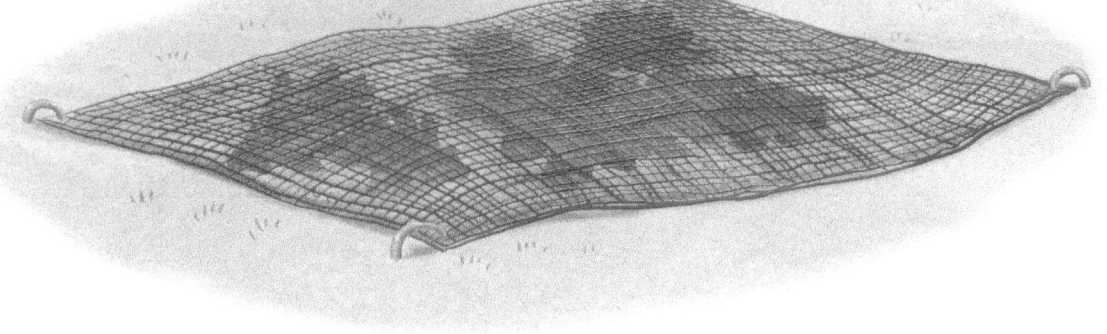

Intelligent insects

Deadly meals

Burying Beetles have a strange idea of baby food. The male and female beetles sniff out a dead mouse or bird. They then spend several hours burying it, slowly rolling its body into a ball of meat. The female lays her eggs in a small hole inside the body, where the eggs hatch into **larvae**. The adult beetles eat the meat and then vomit it back up, like a meaty soup, for the larvae to eat, until the larvae can find food for themselves. If the dead animal isn't quite big enough to feed all the larvae that hatch, the female just eats some of the larvae too!

Burying Beetles get their name from burying dead animals for food.

Mexican jumping beans really can jump! They are plant seeds into which a moth caterpillar has laid her eggs. The larvae hatch, eat the insides of the bean and so grow inside them. When they get too warm, they wriggle around to find some shade, making the beans jump. Bet your baked beans can't do that!

The moth larvae wriggle by snapping their bodies which makes the beans jump around.

Pushy parasites

Parasites are animals that live on other animals. Every animal, including you, has parasites living on or in their body. We can have millions of them, from tiny eyelash mites to tape worms that can grow several metres long in our intestines. But don't worry, almost all parasites are harmless and some can even be quite helpful.

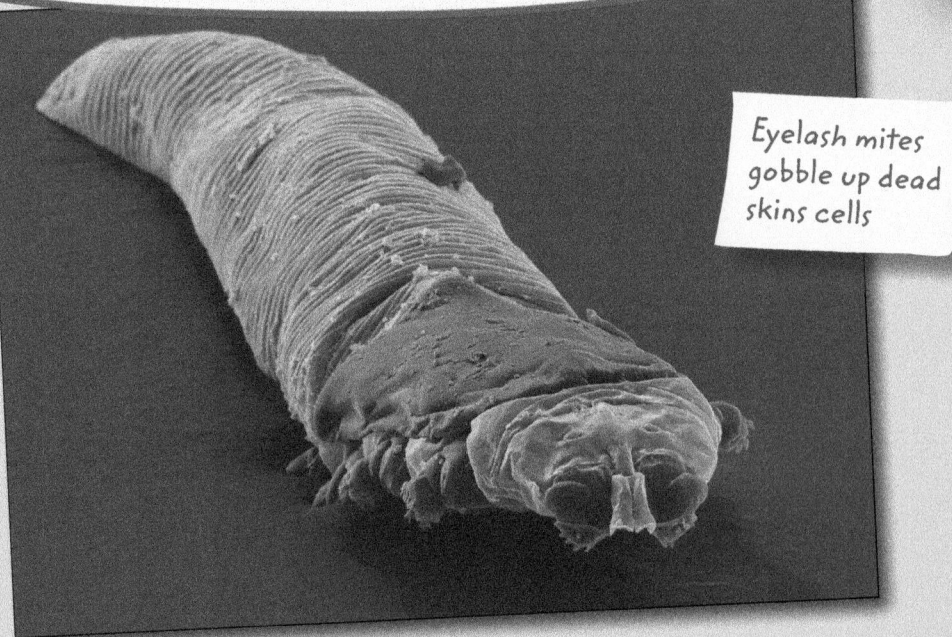

Eyelash mites gobble up dead skins cells

Daily chronicle

September 5th, 1991.

DOCTORS REMOVE 11-METRE TAPEWORM FROM WOMAN'S BODY.

The worm, the longest ever recorded in a human host, was pulled out through her mouth!

Body snatchers

When a parasitic wasp wants to lay her eggs, she goes off in search of a big, fat, juicy caterpillar and injects her eggs under its skin. The eggs hatch into larvae that feed on the caterpillar's blood. When their teeth are sharp enough, they start munching their way out of its body! They release special chemicals that paralyse the caterpillar so it can't crawl away. Up to 60 wasp larvae can be growing inside one caterpillar – amazing!

Wasp larvae eat their way out of a caterpillar's body.

Flour beetles

Some insects, including butterflies, ladybirds, bees and wasps, go through four stages in their life cycle.

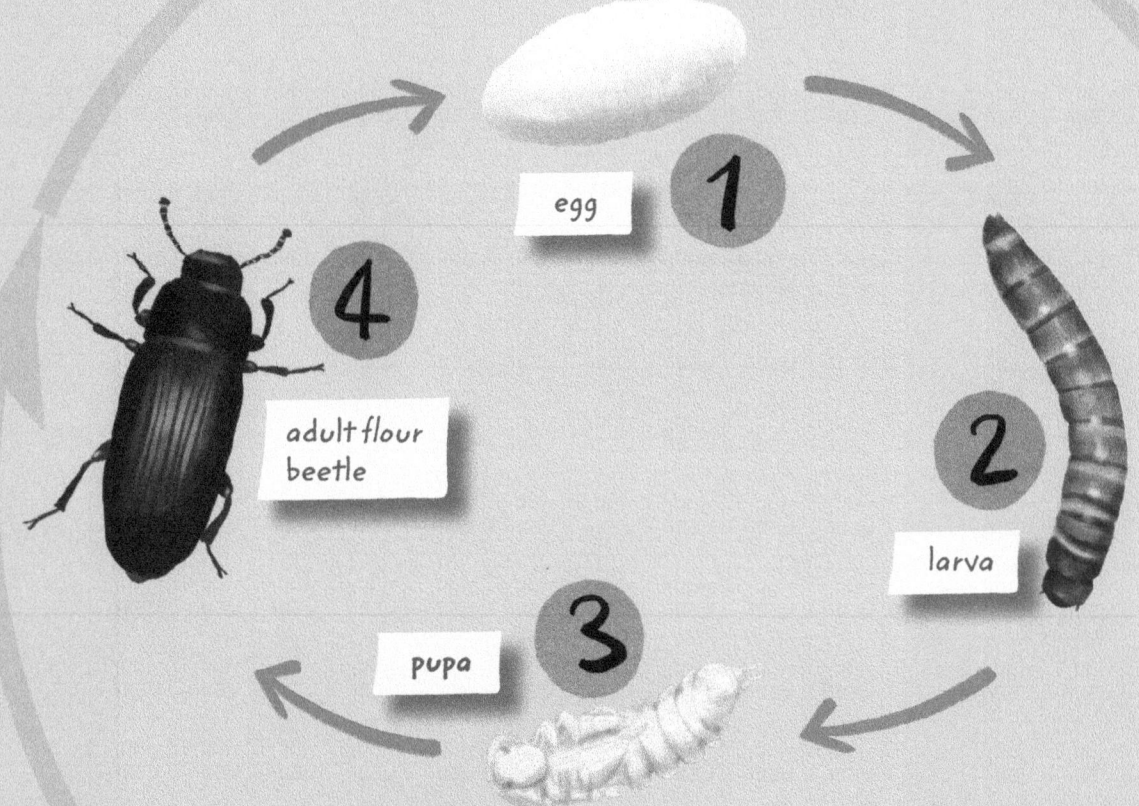

egg

1

4

adult flour beetle

2

larva

3

pupa

Getting in a mess

Flour beetle larvae have a messy habit of shedding layers of old skin as they grow. The larvae shed up to 20 of these skins before they turn into **pupae**. About two weeks later, the pupae split open and the adult beetles crawl out.

Try this!

Want to try hatching your own larvae?

What you need:

- a jam jar
- a spoon
- some bran
- an apple or potato
- some flour beetle larvae or mealworms

What to do:

1 Spoon some bran into the jar.

2 Place a piece of apple or potato on top.

3 Carefully spoon in the larvae or mealworms.

4 Keep the jar indoors.

5 Record when you see **exoskeletons** (which look like empty dry skins) shed on the bran.

6 Record when you see pupae split open to reveal adult beetles.

Hint: renew food before it gets dry or mouldy. Release beetles outdoors as soon as they appear.

Creepy crawlies

Beetle burglars

Some insects have three stages in their life cycle: egg, **larva** and adult. Oil beetles, however, have a sneaky trick. The female lays up to a thousand eggs in a burrow and, as they hatch, the larvae feed on plants nearby. Then they lie in wait, inside a flower, until an unsuspecting mining bee passes. They hitch a ride on the bee's back and are taken to the bee's nest, where they boldly eat the bee's own egg. They also eat the pollen and nectar the bee had stored away to feed its own larvae. Cheeky or what?

larvae

Oil beetle larvae hitch a ride on a mining bee's back.

Three guesses how dracula ants get their name! These tropical ants hunt centipedes, attacking them in swarms and dragging their remains back to the nest. The adults chew the centipedes up as food for their larvae. The adults feed on the larvae's blood, like vampires! They drink just enough blood not to kill the larvae. Parents! What are they like?

The ants work as a team to carry food for their larvae back to the nest.

Hungry spiders

Deadly dancing

An army of over a hundred baby spiders eat their mother then climb onto her web and dance! Sounds like something from a horror movie? It's all part of the life cycle of the black lace-weaver spider. When her eggs have hatched into spiderlings, the female spider carries on laying eggs for them to eat. Then she encourages the spiderlings to snuggle up and eat her alive! Once they've killed her, the spiderlings go off hunting in packs. Together, they can kill **prey** up to 20 times their own size. Scary!

Spiderlings make a meal of mum before dancing on her web.

Strange ...

One type of spider builds life-size models of itself. It sticks them on its web to trick **predators**.

... but true

Decoys stuck on a spider's web can trick predators like birds.

Males and females

Female wolf spiders sometimes eat their mates, but now and again the males get their own back and eat the females first!

Poisonous reptiles

New skins

Some snakes lay eggs but others, like the poisonous
rattlesnake, give birth to live baby snakes or snakelets.
The snakelets are even more dangerous and deadly than
the adults because they tend to squirt out more poison!
But they have to wait for their rattle – the part of their
tail that rattles when they are in danger. They grow it bit
by bit each time they shed their skin. Shedding skin is a
snake's way of replacing an old skin especially when it is
young and growing, and it also gets rid of any parasites!

Rattlesnakes
get their name
from their tail
buttons that rattle
together to warn
off predators.

Strange ...

The regal horned lizard can squirt blood out of its eyes to put off predators.

... but true

Regal horned lizards live in hot, sandy **habitats.**

Many animals renew their skins by shedding them. Even you are shedding thousands of dead skin cells every hour! But some lizards go one step further. They can just separate off their tail if a predator grabs hold of it and then grow a new one later.*

*Remember the lobsters that grew new shells on page 4.

Tadpole tweenies

Is it a fish? Is it a frog? No, it's a tadpole – something between a fish and a frog. Tadpoles grow bigger eating pondweed and sometimes each other!

frogspawn

21 days – tadpoles hatch.

5 weeks – tadpoles start to grow back legs then front legs tails get shorter

11 weeks – tadpoles turn into frogs. Frogs go off to find a mate, and start the life cycle all over again.

Strange ...

These **amphibians**, called caecilians, bring a whole new meaning to the words yummy mummy! When she has babies, the mother grows thick layers of skin then lets her babies crawl over her so they can peel it off and eat it! The babies even grow special teeth to help them chew it.*

... but true

These worm-like amphibians munch on mum in the nest.

*If you think that's going too far, check out the spiders that gobble up their mums on pages 16–17.

Try this!

Do you want to raise some of your very own tadpoles?

What you need:

- an indoor aquarium (large enough for 10–20 tadpoles)
- some water from a pond
- some algae from a pond
- two thermometers
- some frogspawn from a pond

What to do:

1 Half-fill the aquarium with pond water and algae. Place the thermometer so that half of it sits in the water.

2 Carefully transfer a small amount of frogspawn from the pond into the indoor aquarium.

3 Record the temperature in the outdoor and indoor habitats once a day.

4 Record daily changes in the tadpoles.

5 Use your data to decide if tadpoles grow faster in warm or cool conditions.

Hints: take only a small amount of frog spawn.
- Refresh pond water to keep it clean but leave tadpole poo at the bottom of the aquarium because they eat it!
- Remove any tadpoles that die.
- When they grow legs, provide small mealworms as food and renew daily.
- Make sure the tadpoles are looked after at the weekend.
- Return froglets to pond as soon as they develop.

Sneaky birds

Nest thieves

Birds have some sneaky tricks when it comes to parenting. Cuckoos just let others do it for them! When the female is ready to lay her egg, she finds a nest of a smaller bird like a reed warbler in which to lay it. She even copies the colour and pattern of the other bird's eggs to trick the poor warblers into hatching and raising the cuckoo chick as their own. The newly-hatched cuckoo may even pull the other eggs or chicks onto its back and throw them out of the nest, so it can eat all the food the parents bring back!

The bigger cuckoo chick pushes other chicks out of the nest.

Strange ...

Male swallows sometimes throw eggs or chicks out of a nest so that the female who laid those eggs will lay another batch of eggs fathered by them.

... but true

Birds have some of the weirdest ways of attracting a mate. The male frigate bird blows up his throat sac into a giant red balloon to attract a female. He then covers the female's eyes with his wings in case she spots a red balloon she likes better!

Attracting a mate is a key part in an animal's life cycle.

Fearsome fish

Shark pups

What's the scariest thing about great white sharks? Is it that they have 3000 razor sharp teeth growing in rows? Or that they can grow up to seven metres long? That's nothing compared to the fact that the pups eat their own brothers and sisters! The eggs hatch and grow into pups inside their mother's body. They don't have to look far for food – they just eat any eggs that haven't hatched or even smaller and weaker pups!

A Great White shark grows new teeth when the old ones get broken or worn away. Spare teeth grow in rows behind the main set and twist into place when needed.

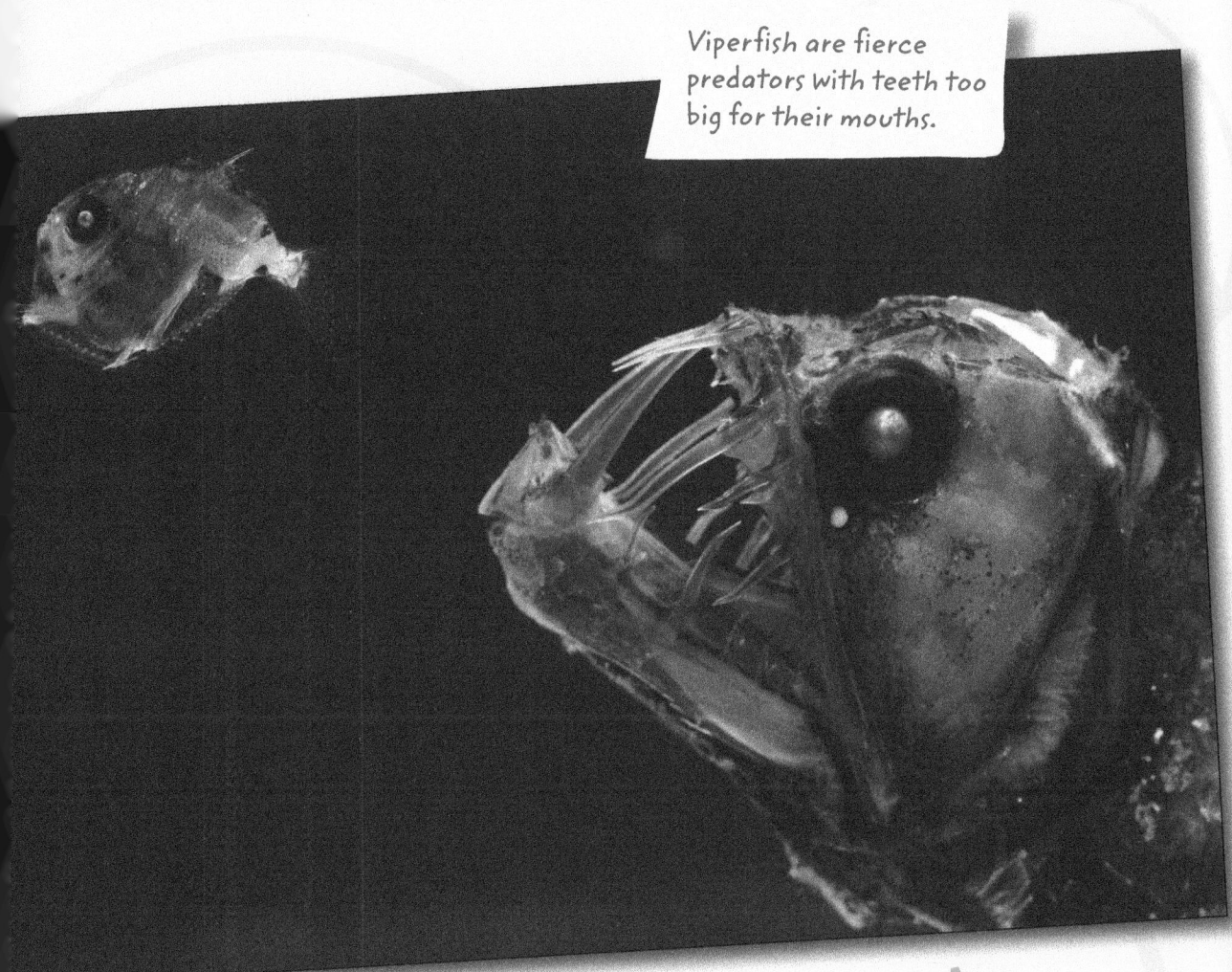

Viperfish are fierce predators with teeth too big for their mouths.

Strange...

Viperfish have a curious lifecycle. The eggs hatch into tiny larvae that have eyeballs that dangle behind them on stalks, looking for prey. As they develop, they slowly pull their eyeballs back into their skulls, whilst growing a fearsome set of teeth for hunting!

... but true

Crafty crustaceans

Spare parts

Lobsters have a neat trick. If they lose or injure a limb, they just grow another one. Do you wish you could do that? However, it's hard growing up as a lobster. Females carry thousands of eggs which hatch as larvae and float on the sea. The few that are not eaten by predators settle on the seabed where they grow into baby lobsters. Out of around 50,000 eggs, only about two will make it as full-size adult lobsters!

Lobsters are crustaceans which means they have a hard shell and live in water. Woodlice are the only crustaceans to live on land.

Try this!

What sort of habitat do you think woodlice prefer?

What you need:

- some card
- some scissors
- a shoebox
- a stopwatch
- two pieces of paper towel, one dry and one damp
- 20 woodlice

What to do:

1 Make a small hole in the bottom of the card.

2 Place it in the shoebox to divide it in half.

3 Put the dry paper towel in one half and the damp paper towel in the other.

4 Place ten woodlice in each side and put the lid on.

Hint: handle woodlice carefully. You can sometimes find them under logs or leaves. Return them outdoors as soon as you have finished.

5 Use a stopwatch to record every minute for ten minutes to see how many woodlice are in each side.

6 Record your data on a chart like the one below.

	0:00	1:00	2:00	3:00	4:00	5:00	6:00	7:00	8:00	9:00	10:00 min.
Dry	10										
Damp	10										

Motherly mammals

Rule breakers

Mammals are warm-blooded animals that have fur and give birth to babies rather than laying eggs. Well, mostly. The duck-billed platypus is a mammal, but it lays eggs. Hedgehogs are mammals, but try stroking their spines! Baby hedgehogs, called hoglets, grow their spines soon after they are born. The spines are soft at first then get more prickly.

Hoglets are born blind and stay in the nest for a few weeks drinking their mother's milk.

Getting a good sleep

Imagine going to bed in November and not waking up until March! Hedgehogs do, because hibernation through the winter months is part of their life cycle. Other mammals that hibernate include bats, brown bears, polar bears and dormice.

Like all mammals, hedgehog mothers feed and care for their young, but if the nest is disturbed, she may just eat them! It is her way of recycling all the energy that went into having her babies. And that's if they don't get eaten by the father first, which can happen if they are boys!

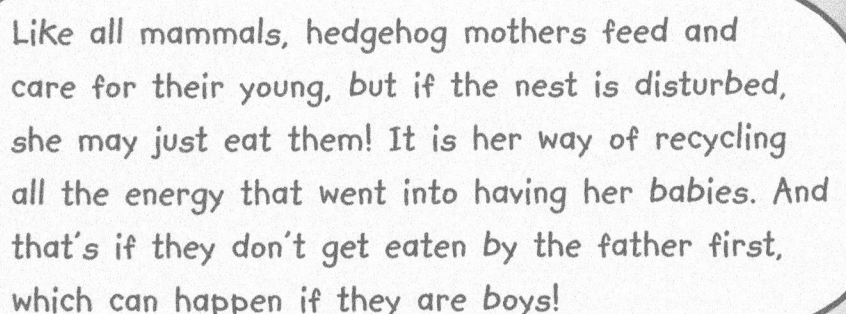

Marsupial mammals like kangaroos give birth to tiny babies or 'joeys' (about the size of a jelly baby). They live in their pouches for about eight months before they hop out.

Human cycles

Babies and birth

So what about your life cycle? Human mothers, like all mammals, give birth to their babies and feed them on milk. They have to wait nine months for their babies to be born so how does that compare with other animals? Female Asian elephants have to be really patient. Their pregnancies last for nearly two years! But at the other end of the scale, mice only carry their babies for a few weeks.

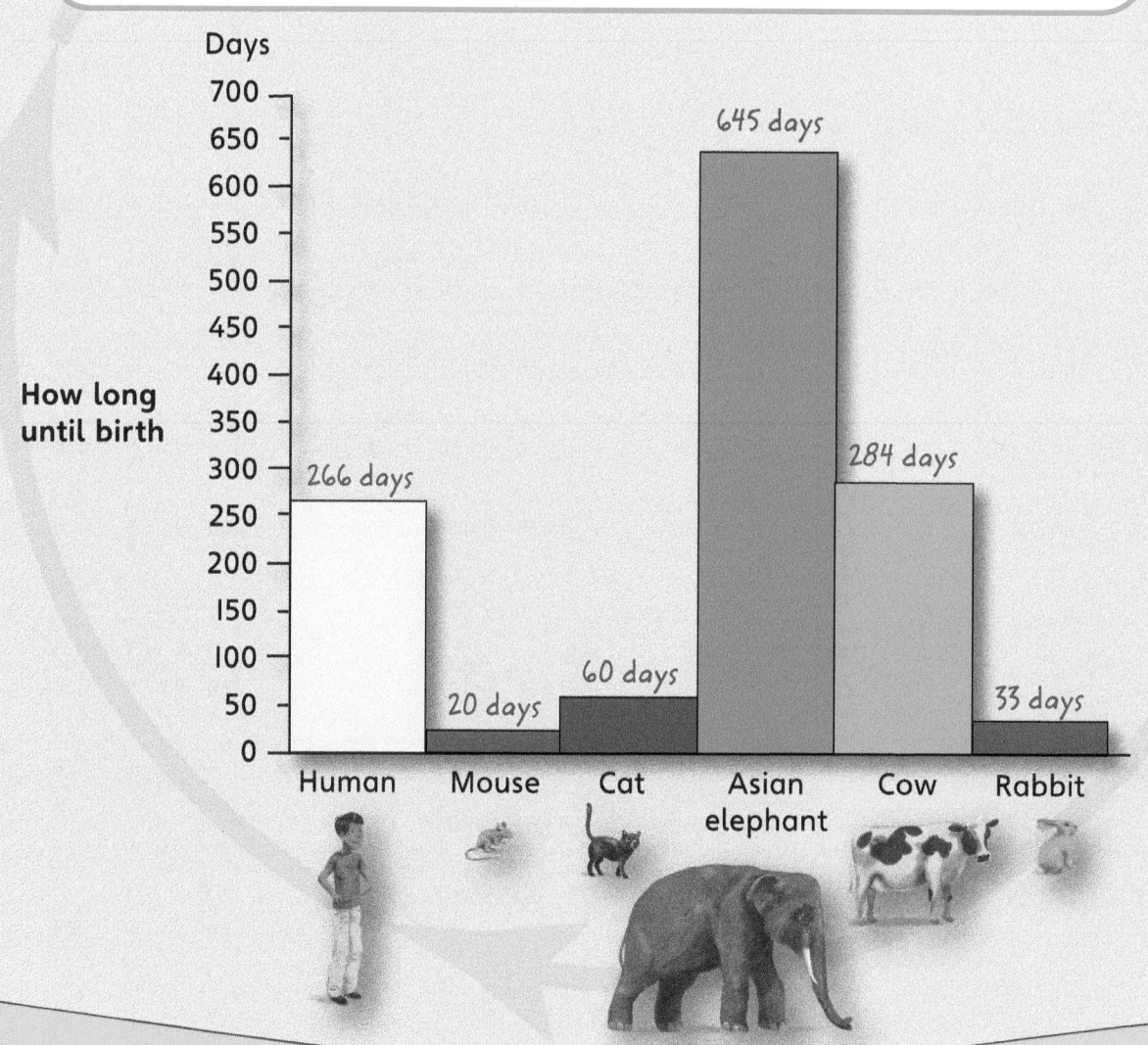

Days

How long until birth

266 days — Human
20 days — Mouse
60 days — Cat
645 days — Asian elephant
284 days — Cow
33 days — Rabbit

How does *your* life cycle compare with other animal life cycles?

Egg

Baby

Child

Adolescent

Adult

Elderly adult

Human *babies* are pretty helpless when they are born. But if you were a kitten or a hedgehog you'd have been born blind. If you were a mouse or a chipmunk your ears would have been stuck to your head! However, human parents may go on caring for their children for 18 years or longer ... that's some parenting, you have to agree!

Glossary

amphibians cold-blooded animals that live partly on land and partly in water

cells tiny units that make up living things

exoskeletons hard outer shells that may be shed

food chain order in which animals feed on other animals or plants, providing them with energy to grow and live

fungi plant-like growths including mushrooms, mildew and moulds

habitats natural surroundings of plants or animals

larva/larvae (pl) insect at the stage in its life cycle when it begins feeding, before it grows wings

marine living in or near seas or oceans

marsupial group of mammals in which the young are born live and initially grow in mother's pouch

predators animals that feed on others

prey animals hunted by others as food

pupa/pupae (pl) insect at the stage in its life cycle between larva and adult

reproduce have babies

Index